The 5S Playbook

A Step-by-Step Guideline for the Lean Practitioner

The LEAN Playbook Series

PUBLISHED

The 5S Playbook: A Step-by-Step Guideline for the Lean Practitioner
Chris A. Ortiz

FORTHCOMING

The Quick Changeover Playbook: A Step-by-Step Guideline for the Lean Practitioner
Chris A. Ortiz

The TPM Playbook: A Step-by-Step Guideline for the Lean Practitioner
Chris A. Ortiz

The Cell Manufacturing Playbook: A Step-by-Step Guideline for the Lean Practitioner
Chris A. Ortiz

The Kanban Playbook: A Step-by-Step Guideline for the Lean Practitioner
Chris A. Ortiz

The LEAN Playbook Series

The 5S Playbook

A Step-by-Step Guideline for the Lean Practitioner

Chris A. Ortiz

CRC Press
Taylor & Francis Group
Boca Raton   London   New York

CRC Press is an imprint of the
Taylor & Francis Group, an Informa business

A PRODUCTIVITY PRESS BOOK

CRC Press
Taylor & Francis Group
6000 Broken Sound Parkway NW, Suite 300
Boca Raton, FL 33487-2742

© 2016 by Taylor & Francis Group, LLC
CRC Press is an imprint of Taylor & Francis Group, an Informa business

No claim to original U.S. Government works

Printed on acid-free paper
Version Date: 20150520

International Standard Book Number-13: 978-1-4987-3035-8 (Paperback)

Library of Congress Cataloging-in-Publication Data

Ortiz, Chris A.
    The kanban playbook : a step-by-step guideline for the lean practitioner / Chris A. Ortiz.
        pages cm
    Includes index.
    ISBN 978-1-4987-4175-0
    1. Production control. 2. Just-in-time systems. 3. Production management. I. Title.

TS157.O78 2016
658.5--dc23   2015018290

Visit the Taylor & Francis Web site at
http://www.taylorandfrancis.com

and the CRC Press Web site at
http://www.crcpress.com

Contents

How to Use This Playbook

In most cases, a playbook is a spiral-bound notebook that outlines a strategy for a sport or a game. Whether for a football game, a video game, or even a board game, playbooks are all around us and when written properly provide immediate and easily understood direction. Playbooks can also provide general information; then, it is up to the user of the playbook to tailor it to their individual needs.

Playbooks contain pictures, diagrams, quick references, definitions, and often step-by-step illustrations to explain certain parts. You can use playbooks to help you understand the entire game or you can pick and choose to focus on one element. The bottom line is that any playbook should be easy to read and to the point and contain little to no filler information.

The *5S Playbook* is written for the Lean practitioner and facilitator. Like a football coach, a facilitator can use this playbook for quick reference and be able to convey what is needed easily. If for some reason the person leading the actual 5S implementation forgets a "play," the playbook can be referenced.

You can follow page by page and use the playbook to facilitate a 5S implementation or you can go directly to certain topics and use it to help you implement that particular "play."

Looking for more supplemental information or Lean coaching from Chris Ortiz? Go to www.leanplaybooks.com and be able to receive ongoing support and advice on how to use the playbook series for training and implementation.

Introduction

At first glance, the improvement techniques within the Lean philosophy appear to provide a solution to many types of production-related issues. A powerful and effective improvement philosophy, Lean can prevent company failure or launch an organization into world-class operational excellence.

I have been a Lean practitioner for more than 15 years and have been involved in many Lean transformations. It does not matter the industry you work in, the product you produce, and even the processes your company uses to transform something to a finished good, the problems and opportunities you face are the same as those of everyone else. Your company is not "different" or an exception. You, as a Lean practitioner, desire a smoother-running facility, reduced lead times, more capacity, improved productivity, flexible processes, usable floor space, reduced inventory, and so on. Organizations implement Lean to make localized improvements or they can use Lean to transform the entire culture of the business. Regardless of your aspirations and goals for Lean, you and many other companies face another similar situation: getting out of what I call *boardroom Lean* and moving toward implementation.

Have no illusions: Lean is about rolling your sleeves up, getting dirty, and making change. True change comes on the production floor, in the maintenance shop, and in all the other areas of the organization and by implementing the concepts of Lean. Companies often become stuck in endless cycles of training and planning, with no implementation ever happening. This playbook is your guideline for implementation and is written for the pure Lean practitioner looking for a training tool and a guideline that can be used in the work area while improvements are conducted. There is no book, manual, or reference guide that provides color images and detailed step-by-step guidelines on how to properly implement 5S (Sort, Set in Order, Scrub, Standardize, Sustain) and the visual workplace. The implementation of 5S is a manually intensive action, and conducting 5S projects properly takes experience and direction. The *5S Playbook* is not a traditional book, as you can probably see. It is not intended to be read like another Lean business book. The images in this playbook are from real 5S implementations, and I use a combination of short paragraphs and bulleted descriptions to walk you through how to effectively implement 5S.

Little or no time is wasted on high-level theory, although an introductory portion is dedicated to the 8 Wastes and Lean metrics. An understanding of wastes and metrics is needed to fully benefit from this playbook. I am not implying that high-level theory or business strategies lack value; they are highly valuable. This playbook is for implementation, so it will not contain filler information. Chapter 1 gets you started right away and outlines how to use the Sort process in a work area. You will see how to properly conduct the red tag process and create categories for the items that need to be removed. Chapter 1 also gives you recommendations on the final disposition of these unneeded items so they truly leave the company.

Chapter 2 is a longer chapter as it covers the Set-in-Order phase of a 5S implementation. Generally, the most labor intensive of the processes, you will find multiple images of shadow boards, floor markings, location designations, and shelf organization. Different examples are provided of Set in Order to enhance your upcoming implementation. Each implementation is different, and seeing diverse examples will help you decide how you will apply the concept at your organization.

Chapters 3 and 4 are relatively shorter in comparison to Chapter 2. The Scrub phase of 5S is explained not only by illustrating the importance of cleanliness during a 5S implementation but also by showing how to prepare the work area ahead of time. Chapter 4 outlines the concept of Standardization by explaining the importance of consistency in each implementation you conduct. Tape or paint colors, as examples, can have certain definitions, and these meanings should be standard across the company. By following a standard visual workplace guideline, implementation will become smoother and easier to lead as everyone is following the same general guideline.

The importance of Sustaining is illustrated in Chapter 5 through real examples of end-of-day reset procedures, daily walk-throughs, 5S audits, and 5S tracking systems. These systems are critical for maintaining performance gains achieved after a 5S implementation and allow a foundation for continuous improvement.

As a bonus, Chapter 6 is a step-by-step guideline for creating and constructing a shadow board. This great tool shows you the 10 steps necessary for a 5S-compliant shadow board for tools and supplies. Multiple examples are given throughout this manual, and I provide more in Chapter 6.

Remember that this is an implementation guide. It is intended to be a hands-on guideline for your kaizen events. 5S is a powerful improvement tool within Lean; this manual will help you conduct effective and worthwhile implementations.

Good luck.

Chris Ortiz

8 Wastes

As a Lean practitioner and teacher, I know the power of the visual factory and how it can be the platform for more improvements. Visual controls are very Lean, and the concept of visual controls is a major part of Lean manufacturing. Lean manufacturing has been and will always be about waste reduction. Developing, sustaining, and improving on a visual factory will remove or reduce a significant amount of waste. Many of you reading this playbook already understand the concepts of waste and Lean. For those of you just getting started, here is a brief description of each of the 8 Wastes of 5S:

Overproduction
Overprocessing
Waiting
Motion
Transportation
Inventory
Defects
Wasted Potential

Overproduction is the act of making more product than necessary and completing it faster than necessary and before it is needed. Overproduced product takes up floor space, requires handling and storage, and could result in potential quality problems if the lot contains defects.

Overprocessing is the practice of extra steps, rechecking, reverifying, and outperforming work. Overprocessing is often conducted in fabrication departments when sanding, deburring, cleaning, or polishing is overperformed. Machines can also overprocess when they are not properly maintained and simply take more time to produce quality parts.

Waiting occurs when important information, tools, and supplies are not readily available, causing machines and people to be idle. Imbalances in workloads and cycle times between processes can also cause waiting.

Motion is the movement of people in and around the work area to look for tools, parts, information, people, and all necessary items that are not available. If a process contains a high level of motion, lead time increases, and the focus on quality begins to decrease. All necessary items should be organized and placed at the point of use so the worker can focus on the work at hand.

Transportation is the movement of parts and product throughout the facility. Often requiring a forklift, hand truck, or pallet jack, transportation exists when consuming processes are far away from each other and are not visible.

Inventory is a waste when manufacturers tie up too much money by holding excessive levels of raw, work-in-process, and finished goods inventory.

Defects are any quality metric that causes rework, scrap, warranty claims, and rework hours from mistakes made in the factory.

Wasted human potential is the act of not properly utilizing employees to the best of their abilities. People are only as successful as the process they are given to work in. If a process inherently has motion, transportation, overprocessing, overproduction, periods of waiting, and defect creation, then that is exactly what involves people. That is wasted human potential.

A visual factory can help you reduce these eight wastes and by doing so will create a much more productive and profitable company for all.

My hope is that you will read this playbook and not only be inspired but also be able to roll up your sleeves and begin your 5S journey after the last page is read.

Lean Metrics

Introduction

To effectively measure your success with 5S, you need to establish a list of critical shop floor metrics that can be measured and quantified. On the production floor, these metrics are often called key performance indicators (KPIs). 5S is a powerful improvement tool that can have a profound impact on reducing lead times, increasing output, improving productivity, and affecting many other types of KPI. In some cases, the change is dramatic. We recommend the following Lean metrics become part of measuring your overall Lean journey:

- Productivity
- Quality
- Inventory
- Floor Space
- Travel Distance
- Throughput Time

Productivity

Productivity is measured in a variety of different ways. Productivity is improved when products are manufactured with less effort. This reduction in effort is essentially the reduction of waste. 5S is put in place to reduce or eliminate all of the steps and time associated with searching for items in an unorganized work area. Once you have reduced or eliminated the time spent searching for tools and parts, walking long distances, and sifting through endless piles of paperwork, there is more time in the day that can be allocated and focused on performing work. And, in this case, it is the value-added work of making products that should be maximized for higher productivity. 5S essentially clears the "smoke" of confusion in the work area and then provides a work environment that harnesses value-added work.

Now, the same number of people before 5S can produce more work in the same amount of time: Fewer steps + Same number of people = Higher productivity. In most cases, we have seen a minimum of 15% increase in overall output from a work area the day after the proper implementation of 5S. However, we have also witnessed output increases between 30% and 50%.

Quality

Improvements to quality are more of a secondary benefit of 5S. The 5S implementations have an impact on internal quality, such as that of rework, defects, scrap, and rework labor, simply by providing a work area with better focus. Every time a worker leaves his or her work area to search for necessities, the worker loses focus; as this happens over time, mistakes can be made. How many times does your dentist get up and walk around the dental office looking for all of his or her tools? Is the dentist searching through unorganized cabinets? No, critical tools and information are readily available so the dentist can focus on the work. This focus not only reduces mistakes but also increases the capacity of the dental office. Better organization always equates to better focus and quality.

Inventory

A lot of money is tied up in parts, material, and supplies, which is why inventory is reduced and measured as part of your 5S journey. The concept of 5S alone does not have such an impact on inventory like other Lean concepts such as kanban and cell manufacturing, but as you organize the work area, any 5S implementation team can be setting the foundation for future inventory reduction projects.

As the work area is being organized, questions may arise about how to organize shop supplies. There may be challenges for the team in creating home locations for excessive quantities of inventory. Should the company buy such high quantities? Does the inventory take up too much shelf space and hence valuable floor space?

Inventory ties up money, contributes to clutter, takes up floor space, and often are some of the most common physical obstacles in the company. Workers spend time shifting material and inventory around just to locate what they need. Time is lost by dealing with excessive inventory just to get to the items required to perform their work. As you are organizing the work area, you should be considering how you may ultimately start reducing inventory levels to help create a more visual workplace.

Floor Space

Floor space comes at a premium, and you need to start looking at the poor use of floor space as hurting the company's ability to grow. Floor space should be used to perform value-added work that creates revenue for the company. It should not be used to store junk or act as a collector of unneeded items. Renting, leasing, or buying a manufacturing building is one of the highest overhead costs. The production floor is in place to serve one purpose: to build products. Although the factory is used for other items, such as holding inventory, shipping, receiving, maintenance, and so on, the production floor should be effectively utilized for value-added work. Value-added work involves the act of building products or the steps needed to change fit, form, or function of the product you intend to sell. Production lines, equipment, and machines all produce a salable product, and the floor space needed to perform this work should be properly used.

As a company becomes less organized and unneeded items begin to accumulate, more space becomes used for non-value-added items. This creates an increase in waste. Over time, items such as workbenches, garbage cans, chairs, unused equipment, tools, and tables tend to pile up, and valuable production space simply disappears. Rather than reduce waste and improve floor space use, the general approach is to add. Add building space, racks, and shelves, and you want to change your perception of space: better use, fewer non-value-added items, less waste, and less stuff.

5S is a powerful Lean tool that can improve the overall use of floor space in the company. The examples in this playbook illustrate that.

Travel Distance

Here is the best way to view travel distance: The farther there is to go, the longer it is going to take. Long production processes can create a lot of waste and can reduce overall performance. Plus, longer-than-needed processes take up floor space. There are two ways to look at travel distance: the distance people walk and the distance inventory (product) is transported.

Travel distance is connected to overall lead times in a process and the entire factory. When work in progress (WIP) is created above required quantities, it takes up valuable floor space and increases the distances that the production line needs. As travel distance increases, floor space becomes improperly used, workers walk farther distances, and lead times are increased. Wait time between processes also increases, and there is added lead time to maneuver inventory.

When work areas are designed incorrectly, they can create a lot of walking for workers, and as they become cluttered, more time is needed to find essential items for work.

As waste is reduced through 5S, the travel distance of product and workers decreases, making travel distance a good Lean metric.

Throughput Time

Sometimes used in conjunction with measuring travel distance reduction, throughput time is the time it takes the product to flow in the production process. Throughput time has a direct impact on delivery; the longer it takes product to move through the plant, the longer it takes to be delivered. Of course, many variables can extend product lead time, so it is wise to simplify the metric by measuring the time when process 1 grabs raw material to the time it is packaged and ready for shipment. Longer production lines require more workstations, workers, tools, conveyors, supplies, and material, which results in additional cost and WIP as well as extended lead times. A physical reduction in distance equates to less throughput time, allowing an organization to promise more competitive, yet reasonable, delivery dates. As the waste of motion and transportation are reduced through concepts like 5S, overall throughput time is reduced with it.

Improving these key Lean metrics and using them as a measurement of your success will have a profound impact on the overall financial success and long-term growth of the company. One could look at these Lean metrics simply as process metrics because they can be measured at the shop floor level. Production workers need to work in an efficient environment to be successful contributors to optimal cost, quality, and delivery. Each Lean metric improved complements another, and another, and so on. As you become more experienced as a Lean practitioner, your understanding of how these metrics relate to each other will become second nature.

5S Supply List

As you begin your journey, you will need to purchase implementation supplies. Supply selection depends on how each organization will incorporate 5S practices into its culture and process. The list of supplies will change over time as additions are made. However, there are common tools and supplies used in every 5S implementation; it is recommended to start with the following items:

Pegboard

Paint (your choice of colors)

Paint pens (shadow boards)

Peg anchors (variety of styles)

Label maker

Label cartridges

Scissors

Floor tape (various colors)

Clipboards

Box cutters

Tape measure (100 feet)

Tape measure (12 feet)

Velcro tape

Glue gun

Stencils (A–Z and 1–10)

Red tags

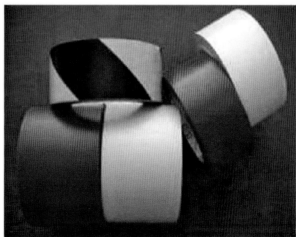

5S Supply Cart

A 5S supply cart is an essential element of your overall 5S and Lean program. It contains all of the essential items for the implementation of 5S and should be stocked prior to the start of any small or large implementation. After every implementation, restock the supply cart so it is ready for the next team. The list I provide is from years of facilitation experience and seems to be a common list from one company to a next. You will modify your list over time, but the basics of the list will not be dramatically changed.

Some companies purchase a cart like you will see in the illustrations, and others simply build one in-house. Keep this supply cart organized and make an inventory list so you can quickly replenish the supplies.

Chapter 1

Sort

Step 1

Sort is the act of removing and discarding all unnecessary items from a work area. Often, it is best to start your 5S implementation in a specific area to help maintain focus and refine your abilities. It can be daunting to think of how you are going to implement 5S across the plant. Focus on one area at a time, break up your plant into "sections," and work section by section like a puzzle. You would be surprised how difficult the sorting function really can be for people. As humans, we love our stuff and struggle to part ways with even the most unnecessary and infrequently used possessions. Just look at the self-storage industry and how it has grown over the years. In a manufacturing environment, items can accumulate quickly and create a mess, hence making things less visible.

Evaluate what is needed to perform the tasks in a work area and remove anything unnecessary. It is good practice to use an identification system while conducting the Sorting step. This system is called red tagging; it can involve the following items:

- Parts
- Tools
- Workbenches
- Garbage cans
- Fixtures and jigs
- Documentation
- Supplies
- Equipment
- Anything that can be removed

Break down the sorted items into three categories:

1. **Garbage and junk (throw away or recycle).** There is not much point in red-tagging garbage, but it still needs to be removed.
2. **Unneeded, never to return.** This category will have most of the red tags. I provide some options of how to deal with these items in further discussion.
3. **Low-use items.** Low-use items are essential but are not used that often, perhaps once a month or in wider time frames. Red-tag these items but place them in a separate pile. Just make sure when you are ready to organize they are organized away from daily-use items.

Make sure to create an inventory list of all items to be removed to help in the final disposition and removal from the company.

The sorting team is responsible for sifting through the workstations to identify all unnecessary tools, supplies, tables, chairs, garbage cans, and so on. It is best to use the production workers on the sorting team who were picked to participate. They have the detailed knowledge of the workstations and can help the other team members. Pair a production worker with someone who does not work in the area. This allows for an outside eye to play "devil's advocate" and question the items in the workstations.

During 5S projects, quick decisions must be made at the sorting phase because the bulk of the work is done during the second phase: Set in Order. This can be an emotional time for the team members as people become close with their work belongings. The collection team is required to be in the red-tag area, which is a temporary staging area for the sorted items. At this point in the process, nothing really has left the building, and it is too early for that to happen. The collection team is receiving everything coming into the red-tag area and organizing things based on the information on the red tag.

Red Tagging

- Organized approach to sorting
- Keeps track of what is being removed
- Allows for efficient removal from facility
- Three parts to a red tag event:
  - Attaching red tags
  - Red-tag area (temporary staging)
  - Removal procedure (covered at the end of this chapter)

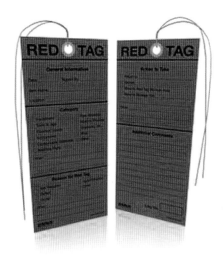 

This is a red-tag area.

Why Sort?

- Unneeded inventory takes up space.
- Extra parts require wasted transportation and motion.
- Unneeded material makes it tougher to find what is needed.
- Unneeded equipment becomes obstacles.
- It is tough to bring in new product lines.
- It is tough to bring in new equipment.
- Sorting clears the confusion that clutter creates.
- Sorting brings clarity on what is truly needed.
- Sorting prepares for Set in Order (why organize items you do not use?).

Is everything here really needed?

Red-Tag Areas

Production

Larger items open up floor space.

Boxes

This is an example of excessive purchasing. These boxes are now obsolete.

Rented Container

Office

- Desks
- Cubicles
- Pencil sharpeners (4) (no pencils anywhere)
- Miscellaneous office supplies

Sorting Hardware

- Remove unneeded hardware
- Make a home for every type of hardware

Body Shop

Placing items on the floor provides a good visual and expedites the sorting process.

Removal Procedure

Once you have completed the Sorting phase, review the items removed and sitting in your red-tag area. The goal of any sorting project is to have all the items removed from the company within 45–60 days. Do not develop a behavior of holding on to items and moving the pile from one part of the plant to another. Develop a removal procedure that can be completed within 60 days.

Removal Ideas

- Donate to local college or trade schools
- Sell items to employees
- Sell items through local and web-based want ads
- Conduct an auction for the public
- Hire an auction house
- Send to sister organizations (if items are needed)

Sorting is an effective way to clear the confusion that clutter creates in a process. You will find that, even after completing only the Sorting part of 5S, it will be easier to work in the work area. However, your 5S implementation has just started, and you have set the tone for the creation of the visual workplace: Set in Order.

Chapter 2

Set in Order

Stage 2

The second stage of 5S is Set in Order. Set in Order is the act of creating locations for all essential items needed in the work area. It is the act of organizing what is needed so it is easily identifiable in a designated place. During this phase, it is recommended that the implementation team work from the floor up, focusing on the layout of the area first.

It is at this point in the implementation when work areas, aisleways, and floor locations are established. Once this is complete, then smaller items such as tools, supplies, fixtures, and parts are given home locations.

Everything has a home.

Painted Work Areas and Aisleways

Painted aisles in maintenance.

Rolled paint works best, especially in dirtier work environments, such as in a maintenance department or production area where there is heavy traffic. Spray paint also works but make sure to purchase an industrial-strength paint.

Tape off line area and paint lines 2 to 4 inches thick.

Floor Tape Work Areas and Aisleways

Black-and-white floor tape at a manufacturing company in Arizona.

Shown is 2-inch yellow floor tape. Clean and painted work surfaces can use floor tape.

Machine Shop

All moveable equipment must have a home location.

Stencils

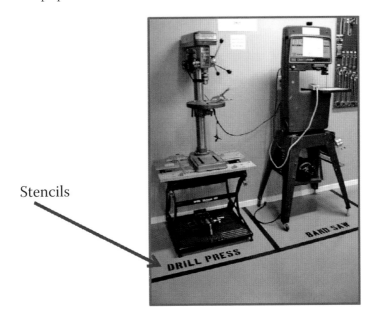

- Provide an address for items.
- Items in rack are labeled URT 7A.

The sign is an address. Note the floor markings.

More floor markings: black-and-white floor tape, Maintenance Department.

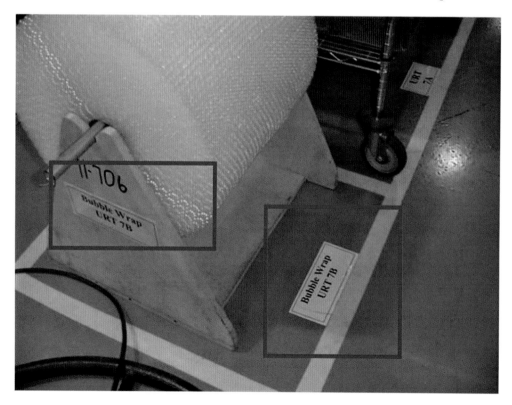

Bubble wrap location and matching labels for proper placement ensures correct items return. Area is clear when items are not in their place.

More floor markings: packaging cart and matching labels.

Maintenance Department used yellow corner markings and stencils.

Service shop floor markings and corresponding color pictures on wall.

Always place "Keep Clear" designations in places where safety is a concern.
In this example, many states have regulatory requirements for the clear distance in front of electrical panels.

Once you have completed the work area outline, aisleways, and locations for items on the floor, progress upward and begin to organize the individual workbenches, equipment, and tool locations. Before you start, do the following preparation steps to help ensure you set up the area properly:

1. Create a list of tools for each station, equipment, workbench, and so on.
2. Separate the pile into common and uncommon use items.
3. Create a list of supplies and other essentials in all workstations.
4. Stage the piles of required items and begin visual organization.

Chapter 6 provides a 10-step illustration on how to create a 5S shadow board.

Tools

These are body shop tools.

These are maintenance tools.

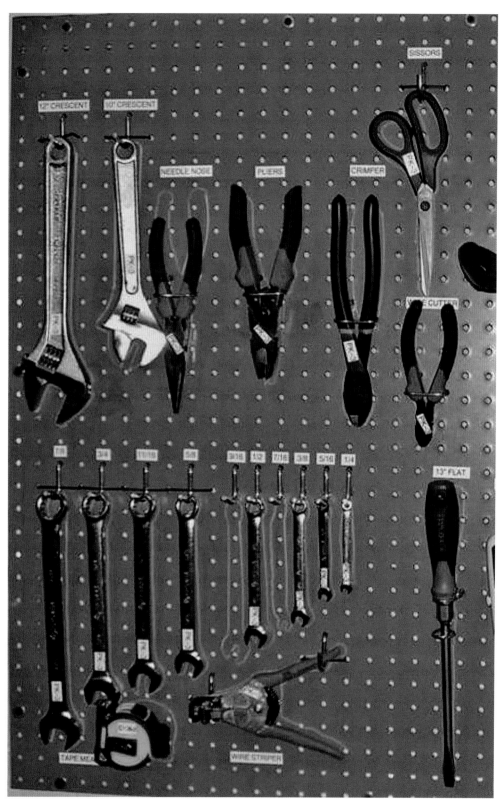

These are shipping tools.

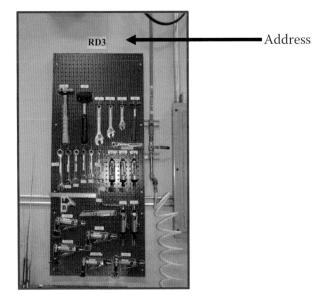

Address

- Critical to tool organization
- Letters represent the department
- Number is the board number
- RD3 placed on every tool

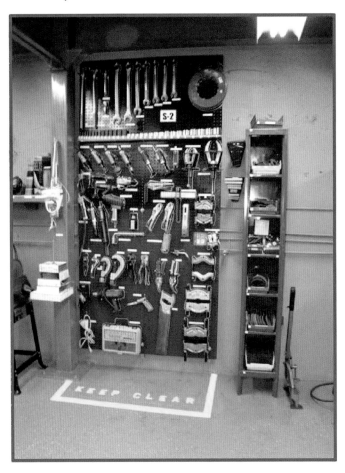

Place "Keep Clear" in front of shadow boards as well.

This is how to label and hang creatively.

Tool Check Cards

Tool check cards are an effective way to keep track of tools in the work area. This generally works best in community areas where a large group of employees shares and borrows tools. It is best to preprint the names of the employees who would access and use the tools the most. If tools are checked out to people outside the department, changeable tool check cards can be made for their short-term use.

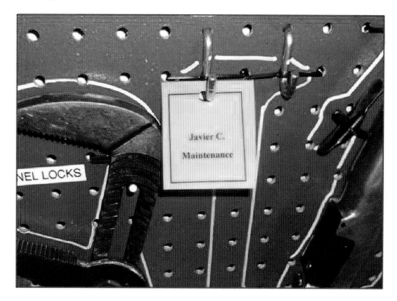

This is a preprinted card for a maintenance technician.

In this example the department color coded the card with yellow to represent maintenance. Other departments can have another color.

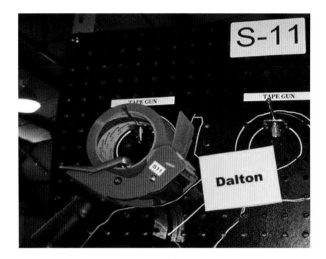

The shipping departments use of check cards make it clear which employee has the tape gun.

Use color-coded cards for different employees; the light blue card is for short-term use of nondepartment employees.

Tool check cards can be made to communicate the status on the tool. Here are some examples:

- Tool broken
- Tool on order
- Tool upgrade

These cards can be made in red or another bright color so they stand out when placed on the shadow board.

Do not forget that tool check cards can be implemented on larger tools, such as power drills, grinders, sanders, and so on. You will find the use of tool check cards will help you keep track of tool use and breakdowns and ensure a reduction in motion and transportation when dealing with tool use. Your tool replacement costs will drop significantly as well.

An effective way to look at the Set-in-order phase is to think of a work area where "everything has a home." Everything deemed necessary to perform the work is clearly marked with designations and locations. This includes items that sit on workbenches and shelves.

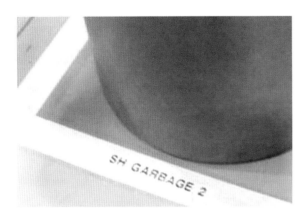 

Garbage can with labels on the can and the floor marking.
Use labels on the garbage can and clearly mark the floor.

Items on shelves have home locations.

Examples

Use corner floor markings and stencils for designating where the trash container is.

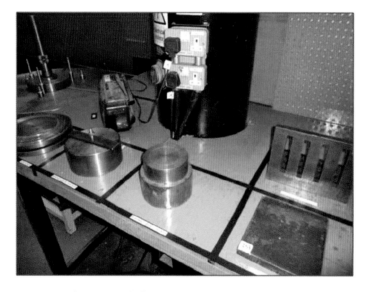

Where inspection tools go is defined by electrical tape as the location markings.

Tool Placement

Ideally, the goal when establishing locations for tools and shadow boards is that the point of use is the best scenario to reduce motion. Use what we call the 5-foot rule: Tools are within a 5-foot reach of the worker. This allows for safe movement of the worker and reduces congestion.

Frequency of use comes into play as well. In areas of high-frequency use, tools should be placed right in the work area at the established 5-foot distance.

There should be no need for a worker to leave the work area for a tool that is used consistently. In some situations, a different approach is needed.

For instance, in a maintenance department there are tools that are used all the time and some tools are used less frequently. Sometimes, tools are used once a month or a quarter. The placement of these tools is not as critical; it makes no sense to place low-use tools in areas of high-use tools.

Often, for these cases two approaches to placement can be used. Review the work area and decide, based on the activities, which tools are best suited for point of use based on frequency. Then, you can create a "community board" where low-use tools are placed and accessed.

Use your best judgment and always try to minimize motion and transportation.

Chapter 3

Scrub

Stage 3

The third stage of a 5S implementation can be done after the Sorting or during the Set-in-Order phases. Scrubbing is the cleaning portion of the implementation. It is not intended to be an intensive clean; here are some areas of focus while cleaning:

- Cleaning and degreasing equipment
- Refill fluids as needed
- Wipe down work surfaces
- Wipe out storage bins
- Wipe shelves and racks
- Clean tools
- Wipe down garbage cans
- Sweep and mop the floor

Cleaning the floor.

In many cases, it is good practice to paint equipment, tables, dollies, shelves, and even the outside of garbage cans during the Scrubbing portion of the project. It creates a showroom appearance and will also help make labels, stencils, and other designations stand out.

Painting hand trucks.

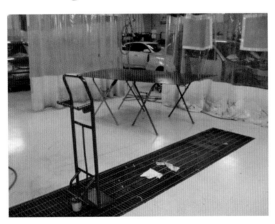

Before machine shop.   After cleaning and painting.

Chapter 4

Standardize

The concept of 5S Standardization is similar to how roadways, highways, and all the visual markings are implemented and used in our everyday lives. As an example, the design of a stop sign is standard across all roadways in the United States.

The design and meaning of all visual roadway systems are identical to reduce confusion. It is recommended that your company come up with a 5S Standardization guideline for your implementation teams.

- Creates consistency in your 5S implementation
- Tool board consistency:
  - All tool boards are painted the same color.
  - Boards have shadows and tool labels.
  - Boards are identified with an address (M1, R3, etc.).
- Identify floor tape or paint colors to define categories:
  - Garbage cans **(green)**
  - Fixed items **(yellow)**
  - Finished goods **(black)**
  - Part locations **(blue)**

Chapter 5

Sustain

Once your implementation of the first four of the five Ss is complete, you need to create a sustaining program that makes sense for your company and your culture. Every company is different, as is how each one establishes the guidelines and practices needed to sustain the improvements. Some companies can rely on the culture with no real management systems in place. Others need formalized systems. Sustaining the 5S program is the hardest. Your sustaining efforts will never end, including continually improving on what was already implemented.

Each company must find its way with Sustaining. Here are a few recommendations:

- Create an end-of-day cleanup procedure
- Conduct a daily/shift walkthrough
- Establish a 5S Audit Sheet
- Create and maintain a 5S Tracking Sheet

End-of-Day Cleanup Procedure

Depending on the area in question, put together a list of items for the area that workers must complete roughly 15 minutes prior to leaving. This cleanup procedure goes beyond simply sweeping the floor and dumping garbage.

Develop a procedure for each department that all shifts follow. Be specific regarding what you want them to do at the end of a shift. Here are some recommendations:

- Empty all garbage and recycle bins
- Sweep the work area
- Return tools to their designated locations

■ Return supplies to their designated locations
■ Place pallet jacks, garbage cans, chairs, and hand trucks in their designated locations

We recommend you post these tasks and allow the operators time to conduct the cleanup to help sustain the improvements made.

Daily Walkthrough

Each area supervisor should take a few minutes after everyone has left to walk through the work area and verify the end-of-day cleanup was completed and all items have been returned to home locations. It is also a good practice to start incorporating the other employees in the work area to become part of the daily walkthrough. Simply create a monthly schedule of who is responsible for each week. His or her job is to perform the walkthrough right after the other staff members have concluded the cleanup.

The purpose of the end-of-day cleanup is to clean and reset the work area for the next shift or the next day with everyone participating in the cleanup. A sense of unity is created. The person conducting the walkthrough is not looking for mistakes from the rest but acting as a backup if something is not returned to its location. This person is aware that when his or her walkthrough week is complete, the following week someone else will "have his or her back." If your company implements 5S to the detailed level described in this playbook, the supervisor walkthrough should be quick. Any small deviations from 5S can be quickly resolved during that walkthrough. This Sustaining approach may be enough to help maintain the area and maximize any improvements to performance.

5S Audit Form/Criteria

In some circumstances, the end-of-day cleanup and daily walkthrough may not be enough. Leaders of the company must still step in if the 5S implementation begins to slip. No system is perfect, and each culture will react differently to the changes and the requirements for Sustaining. We often recommend that in a company's first year of its 5S journey it develop and conduct a 5S audit system to help emphasize the importance of 5S and the commitment from the leaders of the organization. This is not to imply that the audit program will remain a permanent fixture in the company. Never get into a behavior of managing your 5S journey, which is anti-Lean. However, in the beginning, your culture may need some additional structure in the Sustaining phase, and you can decide

when to return to a less-formal system of simple cleanups and walkthroughs.
Here are the basic fundamentals of a 5S audit system.

- 5S Audit Form
- Schedule of audits
- Auditor rotation
- 5S Tracking Sheet

Sample 5S Audit Form: Form 5-1

5S Audit Sheet		
Department/Area:		
Audit Date:	#of Yeses /16= %	
Auditors:		

	Yes	No
Sort (Remove All Unnecessary Items)		
1. Work Station and/or Area is Clear of all Non-Production Required Material		
2. Unnecessary Equipment has been Removed from the Area		
3. Excess and Obsolete Inventory has been removed		
Set in Order (Organize)		
4. Are Cablin and Air Lines Routed Neatly?		
5. All Tools are Organized with Identifications and Home Locations		
6. Locations and Containers for Items, Parts and Supplies are Clearly Marked		
7. Items on Floor are Labelled and Marked		
8. Fixtures are Placed in Proper Locations		
9. Garbage Cans are in Proper Locations		
Scrub (Clean)		
10. Floors, Work Surfaces, Equipment, and Storage Areas are Clean		
11. Garbage and Recyclables are Collected and Disposed of properly		
Standardize (Tasks)		
12. Tool Boards are Consistant in Their Organization and Appearance		
13. Colors on the Floor are Standardized (i.e. Green is for Garbage)		
Sustain (Keep it Up)		
14. 5S Audits are Conducted Weekly and Results are Posted		
15. Auditors are on a Rotation Schedule		
16. End of Clean Procedure is Posted		

Green = 81% to 100% — Area is 5S Compliance

Yellow = 66% to 80% — Area meets minimal standards

Red = 0% to 65% — Area needs immediate attention

Assigning Auditors and Rotation

Once the 5S audit form is designed and ready for use, create a list of auditors who will rotate in and out each week, month, or whatever rotation makes sense. Some companies conduct a 5S audit every week, with one person performing the audit the first week; the following week, another person performs the weekly audit. Other companies may conduct the audits once or twice a month. It depends on your culture and the level of management needed. Start with a weekly rotation and pick a diverse group of employees from all areas of the company to be auditors. Make sure to change your rotation at least every 3 months and ensure that every employee is trained and experienced in auditing to your new 5S program guidelines.

Auditing is a tricky process, and you can expect some backlash. The concept of auditing has historically bothered people, so you need to communicate as a Lean practitioner that the intention of the audits is to find how the company can improve as a whole.

Sample Auditor Rotation

Machine Shop Weekly 5S Audit Rotation Schedule			
Week 1	**Week 2**	**Week 3**	**Week 4**
Kyle Peterson	Candice Tillman	Randy Fox	Levi Handler
Completed Y/N	Completed Y/N	Completed Y/N	Completed Y/N

5S Tracking Sheet

A 5S Tracking Sheet can be a useful tool in communicating the status of the 5S audits and identifying which areas need more attention in Sustaining. It can also act to encourage healthy competition between departments.

Sample 5S Tracking Sheet

5S Tracking Sheet

Area	1st Quarter				2nd Quarter				3rd Quarter				4th Quarter				2015
	Jan.	Feb.	Mar.	*Qtr Avg.*	Apr.	May	June	*Qtr Avg.*	July	Aug	Sept	*Qtr Avg.*	Oct	Nov	Dec	*Qtr Avg.*	*Annual Avg.*
Line 1	100	100	100	**100**	100	90	90	**100**	100	100	100	**100**	89	89	92	**90**	**95.83**
Line 2	100	90	100	**97**	100	90	100	**97**	100	70	90	**87**	80	90	90	**87**	**91.67**
Line 3	90	90	90	**87**	90	90	90	**87**	90	90	90	**83**	100	100	80	**93**	**90.83**
Line 4	100	100	100	**100**	100	95	95	**95**	92	80	89	**80**	90	89	90	**90**	**93.33**
Line 5	100	100	90	**95**	95	90	95	**90**	100	90	90	**90**	89	90	90	**90**	**93.25**
Line 6	100	90	100	**80**	100	95	100	**100**	92	90	92	**90**	90	100	100	**97**	**95.75**
Line 7	100	100	95	**100**	100	100	100	**100**	92	90	92	**80**	100	92	100	**97**	**96.75**
Line 8	90	100	100	**97**	90	100	90	**90**	80	90	100	**90**	90	100	100	**97**	**94.17**
R/D	100	90	90	**87**	90	60	70	**73**	70	70	80	**73**	92	90	70	**84**	**81.00**
Maint.	90	100	100	**97**	100	100	100	**100**	90	90	100	**93**	90	90	100	**93**	**95.83**
Shipping	100	90	100	**97**	90	90	100	**90**	90	100	100	**97**	100	90	100	**97**	**95.83**
Warehouses	90	90	90	**90**	100	100	95	**100**	90	89	89	**90**	100	88	92	**93**	**92.75**

80%–100% Compliant

70%–79% Minimal Compliance

0%–69% Non-Compliant

Incentive Program

Once your implementation of the first four Ss is complete, you need to create a Sustaining program that makes sense for your company and your culture. Every company is different and how each one establishes the guidelines and practices needed to sustain the improvements gained will carry. Some companies can rely on the culture with no real management systems in place. Others need formalized systems. Sustaining the 5S program is the hardest. It is no different from a culture having to maintain anything new. Your Sustaining efforts will never end, including continually improving on what was already implemented. Each company must find its way with Sustaining. Here are a few recommendations:

■ Create an end-of-day cleanup procedure
■ Conduct a daily/shift walkthrough
■ Establish a 5S Audit Sheet
■ Create and maintain a 5S Tracking Sheet
■ Develop a 5S incentive program

Chapter 6

5S Shadow Board:
How-to Reference Guide

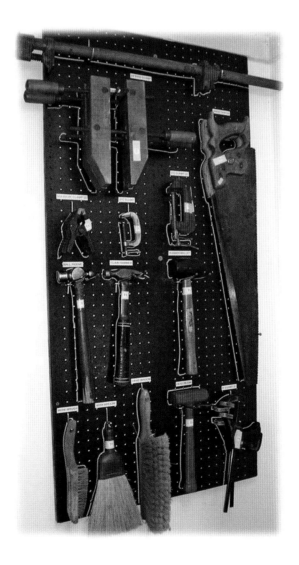

Introduction

■ Shadow boards are one of the most important aspects of 5S and Lean manufacturing. The design and construction of the shadow board should be customized for the area intended for its use.

■ This step-by-step guideline will allow you and your team to successfully build the shadow board and provide comments and suggestions along the way, outlining the importance of each step.

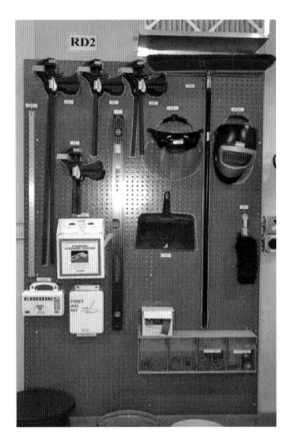
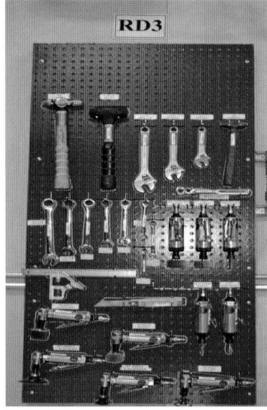

Why Use a 5S Shadow Board?

Tool boards are essentially used to create the visual workplace. Often, tools are placed in a cabinet or toolbox, where tools begin to accumulate, making it difficult to know when tools are missing. Use of tool boards is called "going vertical." The following are the benefits of a tool board:

■ Provides visibility of tools
■ Reduces time needed to locate tools
■ Provides immediate feedback on what is missing
■ Opens floor space as cabinets and tables become unnecessary

How to Build a 5S Shadow Board

1. Paint the peg board; use bright colors
2. Begin to place them on a flat surface verify space requirements
3. Hang board in desired location; use anchors if in walls
4. Place tools and supplies on board with peg hangers
5. Leave enough room for the outline
6. Finalize all tool locations
7. Use a paint pen to outline tools and supplies
8. Make and place labels near item
9. Place colored tape around the tool for a label
10. Designate the board as a location: A1, A2
11. Place a designation label on tools, ideally where it is not touched

Steps 1 and 2

Step 1: Paint Peg Board

- ■ Select bright colors
- ■ Paint the smooth side of the pegboard
- ■ Appearance is important in 5S
- ■ You can also use standard roll paint as well

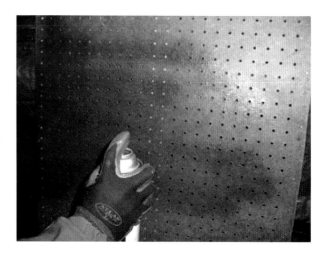

Step 2: Begin to Place the Boards on a Flat Surface to Verify Space Requirements

- ■ Lay out tools
- ■ Allows you to cut the right size peg board

Steps 3 and 4

Step 3: Hang Board in Desired Location

- Walls
- Cabinets
- Side of work benches
- Point of use
- Easy access

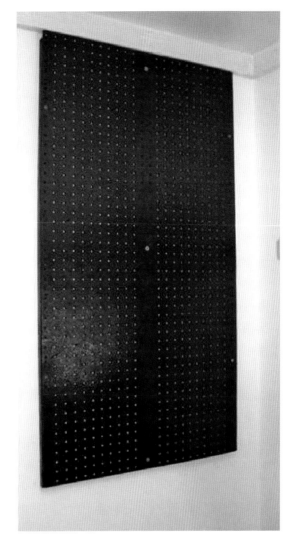

Step 4: Place Tools on Board

- Peg anchors
- Velcro tape
- Fit the tools like a puzzle

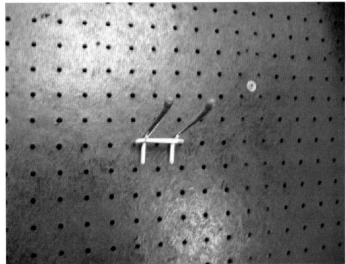

Steps 5, 6, and 7

Step 5: Leave Enough Room for the Shadow or Outline

Step 6: Finalize All Tool Locations

■ Make any changes prior to outlining a tool

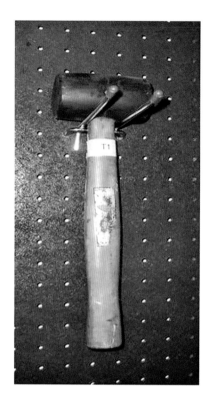

Step 7: Use a Paint Pen to Outline Tool

■ Make sure to draw the tool in its stationary position

 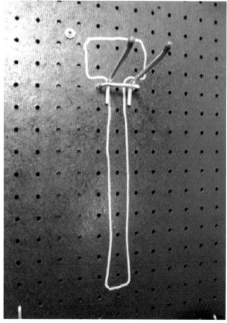

Steps 8 and 9

Step 8: Make and Place Labels Near Tool

Step 9: Place Color Tape Around Tool for Designation

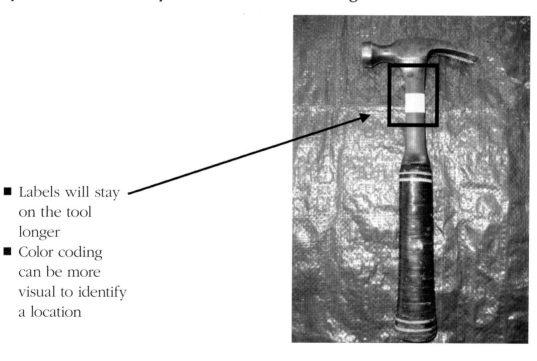

- Labels will stay on the tool longer
- Color coding can be more visual to identify a location

Steps 10 and 11

Step 10: Designate the Board with a Location

- Provides an "address" for tools
- Allows people to find the tool's home

Step 11: Place a Designation Label on Tool

- Place clear tape over the label for security
- Place the label where it is touched the least

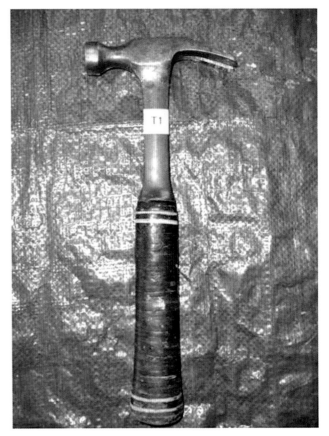

Completion

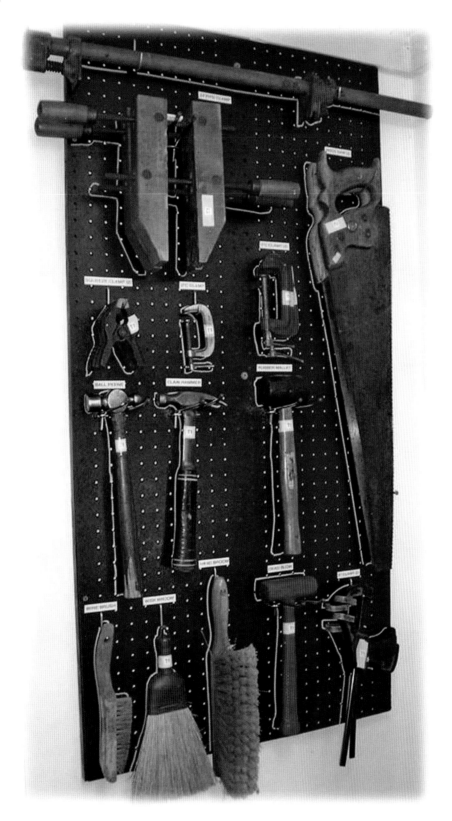

Examples

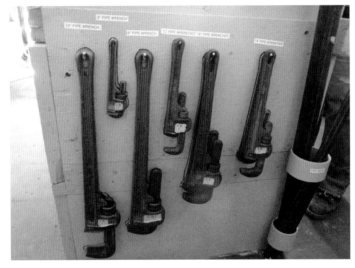

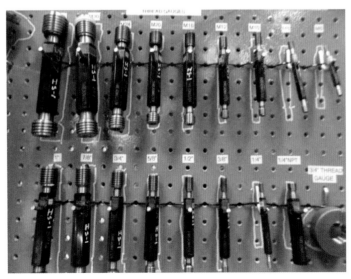

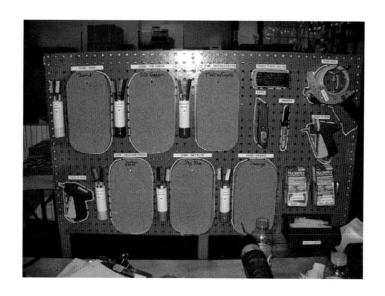

Conclusion

The 5S system is the most visual and tangible element of Lean, and this playbook was written to provide you the most visual and tangible reference there is on the market. 5S is the most universally used Lean concept and is applicable in all work environments. In any Lean implementation, it should be the goal of the practitioner and company to learn the concepts and then tailor them to their needs. Although this playbook provides multiple examples of all the stages of 5S, there are fundamental similarities to each of them. Each company adds its personal touch while maintaining the parameters of what the visual workplace is all about. You will find more opportunity to improve after each and every 5S implementation, through more 5S implementations or through other Lean concepts, such as cell manufacturing, kanban, quick changeover, and total productive maintenance.

5S is a great concept for a start. It helps you learn how to gather and lead a team, identify waste, make tangible changes, see results, and create a buzz around Lean. My hope is that you have learned a lot from this playbook and can use it as a guideline to learn, teach, and implement 5S in your company. Good luck.

Chris Ortiz

Definition of Terms

5S Audit Form: A scoring system used to rate the level of sustaining and as a guideline for continuous improvement.

5S Tracking Sheet: A visual document posted in high-traffic areas that displays the scores from the 5S Audit Sheet.

5S and the Visual Workplace: Lean implementation concept of creating a highly organized work environment where everything has a place. Labels, designations, paint, signage are examples used to create the visual workplace.

Daily Walkthrough: Performed after the end-of-day cleanup, the walkthrough is conducted by a supervisor or worker to verify the cleanup is complete.

Defects: Mistakes made in the process causing rework, material scrap, and lost products.

End-of-Day Cleanup Procedure: A sustaining document that outlines the cleanup and reset requirements for the work area after each shift or day.

Floor Space: Performance measurement of how much factory space is being used to conduct value-added work. Often measured in profit per square foot or revenue per square foot.

Inventory: Higher-than-needed inventory levels due to excessive purchasing of raw material, overproducing work in progress, and unsold finished goods. Inventory ties up working capital, takes up floor space, and adds to longer lead times.

Inventory: The measurement in quantity and cost of raw material, work in progress, and finished goods.

Motion: Movement of workers, generally leaving their work areas to find items unavailable in the work area.

Overprocessing: The act of overperforming work steps, such as redundant effort or extra steps.

Overproduction: The act of producing more product than necessary, performing work in the wrong order, and creating unnecessary inventory.

Productivity: One of the six Lean metrics that is a measurement of a worker's efficiency in a process. Often, it is a comparison of the time allocated to perform work to the actual time the worker took to perform it.

Quality: Internal measurement of rework, scrap, and defects in a production process.

Red Tagging: An organized approach to sorting in which red tags are placed on items to designate them as unnecessary. Red tag items are placed in a staging area for permanent removal from the company.

Right Sizing: Concept of customizing the work area to identify the minimum amount of space needed to store items.

Scrub: Act of cleaning and painting the work area to create a showroom condition.

Set in Order: Act of complete organization of the company by which all items are given home locations.

Shadow Board: A visual mechanism for organizing tools. Shadow boards provide instant feedback on home locations and missing tools and opens floor space by eliminating the need for tool boxes and shelves.

Sort: Act of discarding and removing all unnecessary items from the work area.

Standardize: Act of creating consistency in the 5S implementation through guidelines for the visual workplace.

Sustain: the act of maintaining the work area after a 5S implementation.

Throughput Time: Time associated with all value-added and non-value-added time in a process. It is the time it takes material to get through the first and last steps of the entire factory, from raw material to finished goods.

Transportation: The movement of raw, work in progress, and finished goods throughout the company.

Travel Distance: Measurement of the physical distance of product and workers and the time associated with it. A long travel distance equates to longer lead times in the process.

Waiting: When work comes to a stop due to lack of necessary tools, people, material, information, and parts. Wait time is often called queue time.

Wasted Potential: Poor use of people, including skill sets not being utilized, wrong job placement, and workers consumed in wasteful steps.

Index

About the Author

Chris Ortiz is the founder and president of Kaizen Assembly, a Lean manufacturing training and implementation firm in Bellingham, Washington. Chris has been featured on *CNN Headline News* on the show "Inside Business with Fred Thompson." He is the author of six books on Lean manufacturing (see the list below).

Chris is a frequent presenter and keynote speaker at conferences around North America. He has also been interviewed on KGMI radio and the *American Innovator* and has written numerous articles on Lean manufacturing and business improvement for various regional and national publications.

Kaizen Assembly's clients include industry leaders in aerospace, composites, processing, automotive, rope-manufacturing, restoration equipment, food processing, and fish-processing industries.

Chris is considered to be an expert in the field in Lean manufacturing implementation and has over 15 years of experience in his field of expertise. He is also the author of the following:

Kaizen Assembly: Designing, Constructing, and Managing a Lean Assembly Line (Boca Raton, FL: Taylor and Francis, 2006) (now in its second printing)
Lesson from a Lean Consultant: Avoiding Lean Implementation Failure on the Shop Floor (Upper Saddle River, NJ: Prentice Hall, 2008)
Kaizen and Kaizen Event Implementation (Upper Saddle River, NJ: Prentice Hall, 2009) (translated into Portuguese)
Lean Auto Body (Bellingham, WA: Kaizen Assembly, 2009)
Visual Controls: Applying Visual Management to the Factory (Boca Raton, FL: Taylor and Francis/Productivity Press, 2010)
The Psychology of Lean Improvements: Why Organizations Must Overcome Resistance and Change Culture (Boca Raton, FL: CRC Press and Productivity Press, 2012): Winner of the 2013 Shingo Prize for Operational Excellence in Research